Environmental
IMPACT

What Is the Impact of

EXCESSIVE WASTE AND GARBAGE?

Peggy J. Parks

ReferencePoint Press

San Diego, CA

© 2021 ReferencePoint Press, Inc.
Printed in the United States

For more information, contact:
ReferencePoint Press, Inc.
PO Box 27779
San Diego, CA 92198
www.ReferencePointPress.com

LIBRARY OF CONGRESS CATALOGING-IN-PUBLICATION DATA

Names: Parks, Peggy J., 1951- author.
Title: What is the impact of excessive waste and garbage? / Peggy J. Parks.
Description: San Diego, CA : ReferencePoint Press, 2020. | Series:
    Environmental impact | Includes bibliographical references and index.
Identifiers: LCCN 2019049394 (print) | LCCN 2019049395 (ebook) | ISBN
    9781682828632 (library binding) | ISBN 9781682828649 (ebook)
Subjects: LCSH: Refuse and refuse disposal--Environmental aspects--Juvenile
    literature.
Classification: LCC TD792 .P375 2020 (print) | LCC TD792 (ebook) | DDC
    363.72/85--dc23
LC record available at https://lccn.loc.gov/2019049394
LC ebook record available at https://lccn.loc.gov/2019049395

Contents

Throwaway Societies

The world has a massive problem with the accumulation of waste and garbage, and the problem continues to worsen. As the global population has steadily climbed, from 2.56 billion in 1950 to 7.7 billion in 2019, the waste situation has escalated to what many experts are calling an environmental crisis. According to the World Bank, which works with developing countries to reduce poverty and promote sustainable development, more than 2.2 billion tons (2 billion metric tons) of waste and garbage are produced globally every year. That is an astonishing amount; so much, says *New Internationalist* editor Dinyar Godrej, that it "would circle the planet 24 times if it were piled on to trucks."[1]

The quantity of trash generated by people worldwide is only one part of a larger problem. The other part is how all of that waste is affecting the planet. Although some types of waste break down, or decompose, over time, other types can take years or even centuries. For instance, scientists estimate that it takes at least five hundred years for plastic beverage bottles to break down and up to one thousand years for plastic grocery bags. Glass bottles take even longer—thousands of years—to decompose. When trash builds up in landfills and litters the land and waterways, it poses a serious risk to humans, animals, and their surrounding environment.

An Astounding Array of Disposables

A number of factors contribute to the world's burgeoning waste and garbage. One of the most significant is the widespread use of disposable products. People all over the world use products that are specifically designed to be thrown away after each use. These products are now such a regular part of everyday life that most people who use them could not imagine being without them. Yet until the 1950s these products did not exist.

One of the earliest promotions of disposable products was an August 1955 *Life* magazine article called "Throwaway Living" that hailed the use of disposables as a great convenience. The accompanying photo shows family members who are obviously overjoyed about using products they can just throw away. They are shown amid a floating collection of frozen food containers, paper napkins, paper plates, straws, a disposable diaper, a paper tablecloth, a dispos-able dog bowl, foil pans for cake and pie, garbage bags, drinking cups, and other items that are meant to be tossed into the trash after use.

In the sixty-plus years since the *Life* article was published, use of dis-posable products has soared. These products have evolved from being a novelty to being an essential part of a busy lifestyle. An immense variety of disposables are now sold, including plastic bottles of soda and water, disposable cameras, coffee in single-serve pods, popcorn in microwaveable bags, coffee stirrers, and plastic shopping bags, among a host of others. These and oth-er disposables collectively make up a huge share of the world's waste. According to Earth Policy Institute, 1 trillion plastic grocery bags are used worldwide each year, as are more than 500,000 plastic straws and 500 billion disposable beverage cups.

IMPACT FACTS

Every year, the world produces 2.2 billion tons (2 billion metric tons) of solid waste, which is enough to fill eight hundred thousand Olympic-sized swimming pools.

— Verisk Maplecroft research firm

Projected Growth of Global Waste

All regions of the world are expected to generate increased amounts of waste in the coming decades, with East Asia and the Pacific, South Asia, and sub-Saharan Africa generating the most waste by 2050. The World Bank, which compiled this information, notes that the largest increase will likely occur in sub-Saharan Africa, where the rate of waste generation is expected to triple by 2050. In these regions, much of the waste generated is openly dumped. This, the World Bank warns, could pose serious problems for the environment and for the health and prosperity of the people who live there.

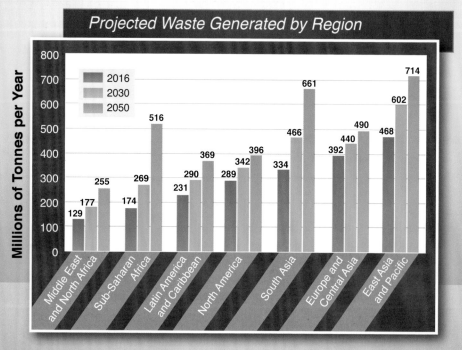

Source: "What a Waste 2.0: A Global Snapshot of Solid Waste Management to 2050," September 20, 2018, World Bank. http://datatopics.worldbank.org.

Another huge contribution to worldwide waste is disposable diapers (called nappies in some countries). In the United States alone, the US Environmental Protection Agency (EPA) estimates that 20 billion disposable diapers are added to landfills each year, creating about 3.5 million tons (3.2 million metric tons) of waste. Because of waste collection services and good landfill management, most of these diapers end up in landfills. But in developing countries, where waste collection services are either rudimentary or nonexistent, disposable diapers often end up tossed into streams and rivers. Inevitably, they end up in the ocean, where they can harm wildlife. "Dirty nappies could be considered the most harmful item of marine litter," says Jo Royle, director of the environmental nonprofit Common Seas. "Science has shown that nappies are causing disease on coral reefs." Also, says Royle, disposable diapers can spread disease. "It is shocking and distressing to see children play on beaches full of dirty diapers."[2]

Worldwide Waste

Although disposable products make up a substantial amount of the world's waste, there are many other contributors as well. One is electronic waste, better known as e-waste. This includes electronic equipment such as televisions, stereos, old VCR and DVD players, photocopiers, computers, printers, and video game consoles, among others. Because of people's fondness for the latest and greatest tech gadgets, electronics such as smartphones, wireless chargers, tablet computers, and e-book readers become obsolete almost as soon as they are purchased, so they are destined to be e-waste too. According to the World Economic Forum, e-waste is growing faster than any other type of waste. More than 53 million tons (48.5 million metric tons) of e-waste were generated during 2018.

Another major contributor to the world's ever-growing waste problem is construction-related debris. This includes waste materials from excavation, roadwork, and demolition, as well as plastics, metal, ceramic, and cardboard. Research has shown that 30 to 35 percent of the total waste generated in developed countries (such as the United States, Great Britain, and the Netherlands) is from building, construction, and demolition activities. In developing countries in Asia, the Middle East, and Africa, construction-related activity represents a much higher percentage of total waste: as much as 90 percent.

Food-related waste also contributes significantly to global trash and garbage. According to the Food and Agriculture Organization of the United Nations, an estimated one-third of the food produced throughout the world for human consumption—about 1.4 billion tons (1.3 billion metric tons)—is either lost or wasted. Lost food refers to food that never makes it to consumers. It either rots in the fields, spills during harvesting or transport, or spoils while in storage. Wasted food refers to food that is discarded by stores that sell groceries or is thrown out by restaurants or consumers. When thrown into landfills, food waste is buried under heavy layers of earth and more waste, and it decomposes without oxygen. This changes the waste into methane and other gases. Scientists say that methane is a powerful greenhouse gas that contributes to climate change, which is the steady rise of global temperatures.

Action Needed

Whether it involves food waste, e-waste, disposables, or construction debris, waste generation is a fact of life—and is a serious problem worldwide. The world's waste has been growing for decades and continues to grow each year. Experts warn that this is a critical situation and a matter of urgency because of threats to the environment. If action is not taken, say the authors of a September 2018 World Bank report, "the world will be on a dangerous path to more waste and overwhelming pollution. Lives, livelihoods, and the environment would pay an even higher price than they are today. . . . What is needed is urgent action at all levels of society. The time for action is now."[3]

Chapter One

The Environmental Impact of Landfills

Tucked into the valley of an isolated mountain range about 20 miles (32.2 km) north of Las Vegas, Nevada, a new mountain grows taller and wider each day. But the mountain is not forming from rocks and soil; it is made of layer upon layer of trash and garbage. This is the site of Apex Regional Landfill, the largest landfill in the United States and one of the largest in the world. The sprawling waste facility, which covers about 2,300 acres (931 ha), takes in about 9,000 tons (8,165 metric tons) of waste every day— equivalent to the weight of twelve hundred fully grown elephants. As trucks rumble into the landfill and dump their massive loads, bulldozers and compactors crush and flatten the waste. This reduces its volume, helps control odors, keeps litter from scattering, and deters rats and other scavenger animals and birds. About once a week, trucks drop a thick layer of dirt over the waste to get the landfill ready for more trash deliveries.

With such a huge amount of waste deposited at Apex every day of the week, every week of the year, it would be logical to assume that the landfill will fill up sometime in the near future. But according to Apex spokesperson Jeremy Walters, this will not happen for a very long time. "Current projections, in terms of capacity out here at Apex, is about 400 years,"[4] says Walters.

More than Garbage Dumps

Along with being America's largest landfill, Apex is also the country's most technologically advanced. It was one of the first landfills to be built in accordance with federal legislation enacted in 1991, which toughened regulations for landfill operations. Before this and earlier environmental laws passed during the 1970s and 1980s, there were virtually no rules about disposal of trash and garbage in the United States. People often burned their trash or took it to a garbage dump—which was basically a gigantic hole in the ground. "Every unwanted article went to the dump," says an article by the city of High Point, North Carolina. "There were old cars, dead animals, bedding, household garbage and even industrial chemicals." The authors of the article continue:

> Fires were a common sight as burning material was dropped into the garbage pile and left to burn. Fires were often started by the operators to make more space in the dump. People often found things in the dump and took them home, risking the spread of diseases. The garbage was rarely covered so rain carried [contaminants] into streams and into the groundwater. Many dumps were located near, or even in water and shoreline areas affecting rivers, bays and estuaries. Odor from the decaying material was overpowering on hot summer days, flies were everywhere and litter from the open piles of garbage blew away from the dump. . . . All dumps had some impact on the environment.[5]

One mandate of the 1991 federal legislation was for all landfills built after October 1993 to have liners on the bottom. The EPA, which enforces US environmental laws, specifies that landfill liners are to be composite, meaning a synthetic material like hard plastic over a thick layer of clay. This forms a barrier to prevent a by-product known as leachate from leaking out of the landfill and contaminating soil and groundwater. Leachate, sometimes called

A bulldozer flattens garbage at the Apex Regional Landfill, located just north of Las Vegas, Nevada. The landfill is the largest and the most technologically advanced in the United States.

"garbage juice," is an inky black liquid that forms when rainwater or melted snow filters through waste in a landfill. As the liquid passes through buried waste, it draws out, or leaches, contaminants from it. Exactly how harmful leachate is depends on what type of trash is contained in a landfill. The liquid's toxicity may range from relatively harmless to dangerously poisonous.

Because of Apex Regional Landfill's desert location, there is very little rainfall, so less leachate forms than in landfills with ample precipitation. But even without rainfall, leachate still forms wherever there is waste because of the natural moisture contained within it. To prevent contamination of soil or groundwater, the federal government requires landfills to have systems in place that collect and remove leachate for treatment and disposal. At Apex, once the leachate reaches the bottom of the landfill, it passes through a drain in the liner and is collected in a tank. In accordance with Nevada environmental laws, the leachate is collected from the tank, treated, and then used to control dust in the landfill.

Birds are a common sight around landfills. Seagulls, magpies, crows, and other birds scavenge the plentiful and stinky waste in search of food. "What we see as a dump, birds see as an endless buffet," says Aleksa Beniusis, who is with the Canadian bird repellant company Lockbird. When birds settle into landfills, it can be harmful to them and to human health.

For birds, the risk is eating garbage that can be deadly to them: plastic, aluminum, drywall, and other potentially toxic waste materials. Also risky for birds is that feeding from a landfill can disrupt their natural migration patterns. The reason birds migrate is to find optimal areas to feed and breed. If they learn that landfills can accommodate their needs, birds may stay rather than migrating. "Landfills are an easy yet dangerous nesting area," says Beniusis.

The risk to humans is that birds and their fecal matter carry numerous diseases, some of which are transmissible to humans—and dangerous. "Birds do not limit themselves to staying directly on the [landfill] sites," says Beniusis. "When they move to other locations, they often carry garbage from the sites with them. As they start spreading this garbage, they are also spreading diseases."

Aleksa Beniusis, "The Environmental Impact of Birds in Landfills," *Waste Advantage Magazine*, October 1, 2018. https://wasteadvantagemag.com.

Leaking Landfills

Apex is often called a model landfill because of its remote desert location, state-of-the-art processes, high-tech equipment, and environmental focus. But not all landfills measure up to such high standards. Although these facilities are regulated by federal and state government agencies and are supposed to be designed so that waste is completely contained, environmental groups say this is not what actually happens. Kirstie Pecci, an attorney for the Conservation Law Foundation, writes: "All landfills leak—some over time and some from day one of operation—leaching toxic chemicals into the ground and the water supply."[6]

Pecci's contention about leaking landfills is supported by the US Geological Survey (USGS). According to a USGS fact sheet, landfill liners and leachate collection systems minimize the risk of leakage but do not always prevent it. Liners can and do fail, and collection systems may not collect all the leachate that escapes. Also, the pipes of these systems require ongoing monitoring and maintenance and can crack, collapse, or fill up with sediment. "All landfills eventually leak into the environment," the authors of the USGS fact sheet write. "Thus, the fate and transport of leachate in the environment from both old and modern landfills is a potentially serious environmental problem."[7]

One landfill that has seriously polluted soil and groundwater is located in Camden, Tennessee, a small town located about 90 miles (145 km) west of Nashville. The three-story-high landfill—which is unlined—was abandoned in 2016 when its owner declared bankruptcy. It is known as "Black Mountain" by residents because of the sludge leaking out of it. A smorgasbord of poisonous waste was dumped in the landfill almost since its opening, including waste from the aluminum, coal, mining, and railroad industries, as well as old diesel fuel from a toxic waste dump. Tests of groundwater and soil in Camden have shown unsafe levels of deadly substances like lead, mercury, and cadmium. The latter is a heavy metal that is so toxic it ranks seventh on the Agency for Toxic Substances and Disease Registry's list of substances that the agency says "pose the most significant potential threat to human health due to their known or suspected toxicity and potential for human exposure."[8] Camden residents and their children have suffered as a result of the toxic, unsafe landfill. Complaints have included headaches and nausea, as well as burning eyes and respiratory problems from the strong ammonia smells emanating from the landfill.

Another leaking, polluting landfill is located in Pasco, Washington. It operated from 1958 to 1993, and during some of those

IMPACT FACTS

The number of US landfills with available space decreased from 7,900 in 1988 to 2,611 in 2019.

—US Environmental Protection Agency

Leachate runs into a collection system. Landfill liners and leachate collection systems minimize the risk of leakage but do not always prevent it.

years the landfill accepted toxic industrial wastes. Studies of the landfill have shown that drums holding industrial solvents, paint sludge, and deadly chemicals known as PCBs are leaking into the soil and contaminating groundwater. Samples have shown a thick layer of these substances floating on the top of the water, which officials say poses a serious threat to human health and the environment.

Landfill Gas

Environmental problems have also been associated with landfill gas, which is another by-product of waste in landfills. The gas forms over time as buried organic matter such as food waste, grass clippings, and paper decomposes under anaerobic conditions, meaning the absence of oxygen. Because oxygen cannot reach the buried matter, rather than changing into humus (or compost), it changes into a gas. Landfill gas is mostly methane. It also contains hundreds of other gases, including carbon dioxide (CO_2), nitrogen, oxygen, ammonia, and hydrogen sulfide. Methane is a greenhouse gas, meaning one that traps and holds heat in the atmosphere, much like a greenhouse traps and holds heat from the sun. In a July 2019 publication, the EPA writes, "Methane is a potent greenhouse gas 28 to 36 times more effective than CO_2 at trapping heat in the atmosphere over a 100-year period."[9] US federal law requires that landfills must collect methane and at the very least burn it off (referred to as flaring). But this is a waste of a potentially useful product, so the EPA recommends that landfill operators turn methane into renewable energy.

Along with being a contributor to climate change, landfill gas can also be a threat to human health. In an October 2019 fact sheet, the New York State Department of Health discusses some of the environmental problems associated with landfill gas. If a landfill does not properly dispose of it, either by flaring or collecting it, the gas can pollute nearby homes and other buildings. "Studies have been conducted in communities near landfills and waste lagoons to evaluate health effects associated with exposure to landfill gases," says the fact sheet. "These studies lasted for several months and reported health complaints which coincided with periods of elevated levels of hydrogen sulfide and landfill odors."[10] These health complaints included eye, throat, and lung irritation; nausea; headaches; sleeping difficulties; chest pain; and aggravation of asthma symptoms. These symptoms are all consistent with exposure to hydrogen sulfide, which is a component of landfill gas.

Gas from a landfill in Volokolamsk, Russia, has caused unbearable living conditions for residents of the town. One of them is Aleksei Stelmakh, who in 2013 built a little house there. Every

IMPACT FACTS

In 2017 more than 282 billion cubic feet (8 billion cu. m) of methane was collected at 370 landfills in the United States and burned to generate 11.5 billion kilowatt hours of electricity, which is about 0.3 percent of total US electricity generation.

— US Energy Information Administration

day he walked onto his porch and reveled in breathing the fresh air and scanning the beautiful nearby forest. Now, however, Stelmakh's view is the gigantic Yadrovo landfill. No more is the air fresh and clean; it is now filled with the rotten egg smell of landfill gas. There is no collection system, so the gas bursts through the soil and vents into the surrounding neighborhoods. Stelmakh has suffered from an increased heart rate as a result of breathing the gas. Children in the area have also suffered from a variety of ailments. Volokolamsk resident Galina Dubrovskaya says her granddaughter has woken up in the night suffering from nausea, dizziness, and headaches. Other children have suffered from a variety of health problems, including nausea and a serious skin condition known as eczema.

Overstuffed Landfills

Because of the potential environmental harm caused by landfills, it is alarming that the volume of trash dumped into them grows larger every year. Although there are many contributors to this waste growth, one of the biggest is America's low recycling rate. According to a July 2018 EPA fact sheet, less than 26 percent of the solid waste generated in the United States during 2015 was recycled. The inevitable result is that tons of recyclable plastic, paper, cardboard, glass, and aluminum are tossed into landfills, adding to the burgeoning heaps of waste and causing landfills to fill up faster than they should.

IMPACT FACTS

Sweden sends less than 1 percent of its waste to landfills.

— Avfall Sverige, Sweden's waste management and recycling association

Complicating the already low recycling rate in the United States is that for years, lower-quality recyclables were shipped to China. And not only the United States did this; Canada, Japan, South Korea, and many European countries also sent their recyclables to China. In 2017 Chinese officials announced that they would no longer accept these recyclables, and the waste began piling up. Com-

Less than 26 percent of the solid waste generated in the United States is recycled. The result is that tons of recyclable plastic, paper, cardboard, glass, and aluminum are tossed into landfills.

munities throughout the United States began dumping their recyclables into landfills, which contributes to shrinkage in available landfill space. According to data from the publication *Waste Business Journal*, total US landfill capacity is expected to decrease by more than 15 percent by 2023. "There's a natural end of life for every single landfill," says Jim Fish, the chief executive officer (CEO) of Waste Management, Inc. "No landfill lasts forever."[11]

Difficulties Faced by Asian Countries

The United States is a huge country with a vast amount of land. So even with shrinking landfill space, there will still be a vast amount of space for decades, even centuries. But some countries are struggling right now with more waste than their landfills can accommodate. One of these countries is Japan. Although the Japanese people generate less than half the waste that Americans create, they still collectively produce more than 40 million tons (36.3 million metric tons) of waste per year. Japan

From a Despised Dump to a Glorious Park

In the New York City borough of Staten Island, a beautiful public park is being created. When it is completed sometime around 2036, the park will be huge—2,200 acres (890 ha), which is nearly triple the size of New York's Central Park. The sprawling park will feature rolling hills, tidal creeks and wetlands, grassy meadows, and fields of colorful wildflowers. Visitors will be able to walk through scenic nature trails, go horseback riding or kayaking, and enjoy spectacular views of the Manhattan skyline. In the winter visitors can enjoy sledding and cross-country skiing and can visit one of the park's restaurants for hot chocolate.

When looking at the work in progress and considering how glorious the park will someday be, it is difficult to believe what the area once was: the largest landfill in the world, despised by everyone in the area. "The biggest garbage dump on the planet once contained 150 million tons of reeking trash," says environmental writer Elizabeth Royte. "No more." The Fresh Kills landfill first opened in 1948 and was serviced by the New York City Department of Sanitation. After more than fifty years, Staten Islanders had grown so sick of the stench that they sued the city to shut the landfill down, and they succeeded. Fresh Kills was permanently closed in 2001. Several years later a community group conceived the idea of an expansive new park on the site of the old landfill. That concept led to Freshkills (now one word) Park.

Elizabeth Royte, "New York's Fresh Kills Landfill Gets an Epic Facelift," *Audubon*, July–August, 2015. www.audubon.org.

is an island nation, and its massive volume of waste is straining available landfill space. Some experts warn that if current waste generation trends continue, all Japan's landfills could be filled up in about twenty years.

Another country that struggles with burgeoning waste production is India. According to Indian environmental officials, the country generates more than 68 million tons (62 million metric tons) of waste each year, of which nearly 6.2 million tons (5.6 million metric tons) is plastic. The landfills are much like America's old dumps—no liners and little government oversight. Methane often catches fire at landfills, causing widespread burning of garbage and noxious air pollution.

In the metropolitan city of Delhi, India, the Ghazipur dump has become so overfilled and huge it is cynically referred to as the "Mount Everest of Trash." As tons of garbage and waste have been dumped into Ghazipur year after year, the mountain of waste has steadily grown higher and higher. It was designed to be a maximum of 67 feet (20 m) high, but this limit was met in 2002—and the landfill still kept accepting waste. This had a disastrous outcome. On September 1, 2017, one of the dump's slopes collapsed, and an avalanche of waste poured into roads and spilled into an adjacent canal. "It was a flood of trash," says Vishal Kumar, a Delhi resident who was caught in the landslide. "I saw heaps of garbage coming down the hill like a flood and suddenly, we were swept into the canal. For a moment, everything went dark."[12]

A Tough Problem to Solve

Countries all over the world have a waste problem, and their burgeoning landfills are irrefutable evidence of that. Some people refer to these monstrous trash mountains as necessarily evils: even though they serve a useful purpose, they are often smelly, leaky, and polluting. Some, like Apex Regional Landfill near Las Vegas, are state-of-the-art waste facilities that are located far away from residential areas and recycle their by-products in responsible ways. Others, like the Ghazipur dump in Delhi, India, are failing and pose a serious environmental hazard. Solutions are needed—but as long as global waste generation continues to grow, what those solutions should be remains unknown. As the World Bank's Ede Ijjasz-Vasquez opines, "There is no end in sight to this trend."[13]

Chapter Two

The Plastic-Plagued Planet

In a June 2018 speech at an environmental conference in China, United Nations resident coordinator Nicholas Rosellini spoke about the many threats to the environment. He offered several examples, one of which was plastic waste. Rosellini talked about the massive quantity of plastic waste that exists, and he cautioned about how this waste threatens the environment. "Plastic waste causes plastic pollution," he said. "If it is burned it releases an incredible amount of toxins into the air. If it is left on the ground or in the sea, it releases toxins. In fact, the most common source of pollution in our oceans comes from plastic and this has already started to severely impact marine life."[14]

From Plastic Products to Plastic Pollution

Scientists and environmental advocates throughout the world share Rosellini's concern about threats from plastic waste. It is a serious problem—one that has come about because of the tremendous amount of plastic that has been produced and continues to be produced. Adding to the problem is that most of this plastic ends up as rubbish rather than being recycled.

This was the key finding of a 2017 study by researchers from the University of California, Santa Barbara; the University of Georgia; and the Sea Education Association. The study, which was published in the journal *Science Advances*, was the first global analysis of the production, use, and outcome of all plastics ever made. The researchers found that as of 2015, more than 9 billion tons (8.2 billion metric tons) of plastic had been manufactured since the early 1950s. Of that, nearly 7 billion tons (6.4 billion metric tons) had already become waste, with only a small percentage of the waste recycled. "If that seems like an incomprehensible quantity, it is," says *National Geographic* journalist Laura Parker. "Even the scientists . . . were horrified by the sheer size of the numbers."[15]

The vast amount of plastic in the environment is a direct result of the mind-boggling array of products that are made of plastic. One example is disposable plastic containers. These containers hold everything from shampoo and liquid laundry detergent to milk, peanut butter, and mustard. Many plastic products are specifically designed to be disposed of after a single use. Environmental scientists say these products are especially problematic, largely because they are so numerous. Dinyar Godrej, a human rights activist and coeditor of *New Internationalist* magazine, writes, "We discard this stuff, often without a second thought, but also because there is so much of it and it just keeps coming."[16] Godrej adds that half of all plastic items produced each year are single-use items. These include plastic grocery bags, straws, coffee stirrers, soda and water bottles, and most food packaging—just to name a few.

IMPACT FACTS

Humans have manufactured 9.1 billion tons (8.3 billion metric tons) of plastic since the 1950s—which equals the weight of roughly 1 billion elephants or 47 million blue whales.

— The Food and Agriculture Organization of the United Nations

Plastic Is Forever

The length of time it takes plastic to decompose is why it is considered one of the most destructive pollutants that exist. As unimaginable as it may seem, all the plastic that has been produced since the 1950s is still around today. The only exception is the relatively small amount of plastic (10 to 12 percent) that has been incinerated. Parker offers a whimsical yet thought-provoking scenario for consideration: "If plastic had been invented when the Pilgrims sailed from Plymouth, England, to North America—and the *Mayflower* had been stocked with bottled water and plastic-wrapped snacks—their plastic trash would likely still be around, four centuries later."[17]

During the 2017 *Science Advances* study, the researchers concluded that if current trends continue, about 13 billion tons (12 billion metric tons) of plastic waste will be in landfills or littering the natural environment by 2050. "Most plastics don't biodegrade in any meaningful sense, so the plastic waste humans have generated could be with us for hundreds or even thousands of years," says Jenna Jambeck, study coauthor and associate professor of engineering at the University of Georgia. "Our estimates underscore the need to think critically about the materials we use and our waste management practices."[18]

The vast amount of plastic waste is a direct result of the mind-boggling array of products that are made out of plastic. One example is disposable plastic containers, which are used for everything from milk to shampoo.

Single-use plastic grocery bags are particularly worrisome because they last for centuries. In the United States alone, 100 billion single-use plastic grocery bags pass through the hands of consumers. "Laid end-to-end," says the Earth Policy institute, "they could circle the equator 1,330 times."[19] Research has shown that the average time that plastic bags are used is only about twelve minutes—yet these bags take up to one thousand years to break down.

Terrible Toll on Wildlife

Plastic is harmful to the natural environment in countless ways, and its impact on wildlife is among the most devastating effects. According to the United Nations, at least eight hundred species worldwide are threatened by marine debris, and up to 80 percent of that debris is plastic. Scientists estimate that as much as 14.3 million tons (13 million metric tons) of plastic ends up in the ocean every year, threatening whales, dolphins, other sea mammals, fish, seabirds, and sea turtles. Research has shown that about one hundred thousand marine animals are strangled, suffocated, or injured in other ways by plastics every year. "The wildlife impact of plastics pollution is staggering,"[20] says investigative journalist Barry Yeoman, who frequently reports on environmental issues.

> **IMPACT FACTS**
>
> Every year 100 billion plastic bags are used by Americans. Tied together, these bags would reach around the earth's equator 773 times.
>
> — Earth Policy Institute

In a June 2019 article, Yeoman describes the countless creatures that are injured or killed every year throughout the world because of plastic in the environment. Referring to lost or abandoned plastic fishing nets (called "ghost nets"), which present one of the greatest threats to marine creatures, he writes, "Victims include more than 340 species, from bottlenose dolphins, humpback whales and endangered Hawaiian monk seals to brown pelicans and every known species of marine turtle."[21]

In March 2019 a young whale that died near Davao City, Philippines, was killed by plastic. The whale was a type known as a Cuvier's beaked whale and was about 15 feet (4.6 m) long. The creature was emaciated and very sick, vomiting blood, and listing to one side when it tried to swim. Because the creature was clearly near death, the agency asked for someone to come and pick up its body. By the time marine mammal expert Darrell Blatchley got there, the whale had died. Blatchley transported it back to his laboratory, where he performed a necropsy—a thorough examination to determine the cause of death.

Before he even cut the whale open, Blatchley was fairly certain that plastic had killed it—but he was shocked at the massive amount of it in the creature's belly. "Plastic was just bursting out of its stomach," he says. "We pulled out the first bag, then the second. By the time we hit 16 rice sacks—on top of the plastic bags, and the snack bags, and big tangles of nylon ropes, you're like—seriously?"[22] In all, Blatchley pulled 88 pounds (40 kg) of tightly jammed plastic waste out of the young whale. He determined the cause of death to be starvation and dehydration. The severe plastic blockage had made it impossible for food and water to travel from the whale's stomach to its intestines. Blatchley sees this kind of tragic occurrence over and over again and finds it heartbreaking as well as terribly frustrating. "It's just tragic that this is becoming the norm," he says, "to expect that these whales will die because of plastic rather than natural causes."[23]

Sea Turtles in Danger

Plastic waste in the ocean also poses a grave threat to sea turtles. This was one of the most alarming findings of a December 2017 worldwide study by researchers at the University of Exeter in the United Kingdom. The study revealed that hundreds, or per-

Plastic Polluting Islands

When determining the largest sources of plastic waste in the ocean, scientists have typically cited certain Asian countries: most notably, China, Indonesia, the Philippines, and Thailand. But digging deeper reveals more of the story. According to a September 2019 *Forbes* article by environmental journalist Daphne Ewing-Chow, research has shown that the Caribbean islands are the biggest plastic polluters per capita in the world. One of these is the small Caribbean island of Saint Lucia, which generates more than four times the amount of plastic waste per person that China produces. In fact, littering is so prevalent in Saint Lucian society that people openly throw trash out of car windows, on the roadsides, in bushes, and into the water. This trash has wound up trapped in drains, where it leads to clogs and flooding; accumulated in rivers, where it affects water quality; and deposited into the ocean, where it harms marine life such as fish and turtles.

Ewing-Chow cites a 2015 scientific study, in which the researchers found that of the top thirty global plastic polluters per capita, ten are in the Caribbean region. These island nations (some of which have dual islands) include Trinidad and Tobago, Antigua and Barbuda, Saint Kitts and Nevis, Guyana, Barbados, Saint Lucia, Bahamas, Grenada, Anguilla, and Aruba. Of these ten, the biggest plastic trash producer per capita was found to be Trinidad and Tobago. This island nation, says Ewing-Chow, "produces a whopping 1.5 kilograms of waste per capita per day—the largest in the world."

Daphne Ewing-Chow, "Caribbean Islands Are the Biggest Plastic Polluters Per Capita in the World," *Forbes*, September 20, 2019. www.forbes.com.

haps even thousands, of sea turtles die every year after becoming entangled in plastic trash. The biggest threat is to hatchlings and young turtles. The researchers found turtles entangled in abandoned or lost fishing nets, plastic twine, nylon fishing line, six-pack rings from canned soda and beer, plastic packaging straps, plastic balloon string, kite string, plastic packaging, and discarded anchor line, as well as other types of trash. Turtles that did not die from being entangled in plastic trash were forced to drag the debris with them as they swam.

Brendan Godley, who teaches conservation science and directs the Centre for Ecology and Conservation at the University of

Exeter, was lead author of the study. He warns that as the plastic pollution crisis grows worse, more and more turtles are likely to become entangled and killed. "Plastic rubbish in the oceans, including lost or discarded fishing gear which is not biodegradable, is a major threat to marine turtles," says Godley. "We found, based on beach strandings, that more than 1000 turtles are dying a year after becoming tangled up, but this is almost certainly a gross underestimate. Young turtles and hatchlings are particularly vulnerable to entanglement."[24] Godley goes on to say that he and the other researchers surveyed experts and learned that entanglement in plastic and other plastic pollution pose a greater threat to the survival of some turtle populations than oil spills.

Endangered Land Animals

Plastic waste in the environment is deadly not only to marine wildlife, but also wildlife species that live on land. A 2018 study by the scientific research group Forschungsverbund Berlin estimated that one-third of all plastic waste ends up in soil or bodies of freshwater. As a result, land animals are also threatened by plastic waste. This includes cows and other farm animals; birds, raccoons, and other wild animals; and domesticated animals such as dogs and cats. Like marine animals, those that live on land can accidentally ingest plastic, which can lead to intestinal blockages and death.

Land animals can also become entangled in plastic in various ways. They may, for instance, get plastic wrap wound around their wings or legs. This makes it difficult for the animal to fly, walk, or run away from a predator or an oncoming vehicle. In addition, being entangled in plastic can make it hard for an animal to move around to get food and water. Some animals have been sliced up

Plastic Pollutes the Great Lakes

In study after study, plastic waste has been revealed as a massive threat to oceans and marine wildlife. But scientists have also learned that the Great Lakes are being harmed by plastic waste. These huge lakes—Lake Superior, Lake Michigan, Lake Huron, Lake Erie, and Lake Ontario—are a series of inter-connected lakes that collectively make up the largest freshwater system in the world. Research has shown that plastic debris makes up about 80 percent of the litter found along the shorelines of the Great Lakes. According to a study by the Rochester Institute of Technology, nearly 22 million pounds (10,000 metric tons) of plastic waste enter the Great Lakes each year—with more than half in Lake Michigan alone.

Of further concern, researchers say that microplastics have been found in all five of the Great Lakes. Just as in the ocean, plastics in the Great Lakes are broken down by wind, sunlight, and waves. From this action, microplastics form. Scientists have observed microplastics in Great Lakes fish and in drinking water and locally brewed beer. In some ways plastic pollution in the Great Lakes represents an even greater problem than in the ocean. This is because plastic in the Great Lakes does not circulate around the globe; it has nowhere to go. And especially alarming is that the effects on human health remain unknown.

Like marine animals, animals that live on land can accidentally ingest plastic, which can lead to intestinal blockages and death.

by plastic ring beverage holders or strangled when plastic loops get around their necks.

Another danger for land animals from plastic waste occurs when an animal such as a raccoon, dog, or cat gets its head stuck in a plastic food container like a peanut butter jar. This can lead to overheating, dehydration, suffocation, starvation, and eventually death. And with a plastic container on their head, these animals are less able to defend themselves against predator animals.

Whirling Catastrophes in the Ocean

Scientists have discovered another disturbing result of plastic waste: garbage patches, or plastic accumulation zones in the world's oceans. These are areas where rotating currents known as gyres are located. The National Oceanic and Atmospheric Administration (NOAA) explains, "You can think of them as big whirlpools that pull objects in." As these whirling gyres suck plastic trash into them, they get larger and larger, forming garbage patches. But the term *patch* is somewhat misleading because it connotes a floating island of trash, and these garbage-filled gyres are much more extensive than anything that only floats on the surface. "Instead," says NOAA, "the debris is spread across the surface of the water and from the surface all the way to the ocean floor. . . . Garbage patches are huge!"[25]

There are five gyres in the world's oceans, and according to the NOAA, each of them has been found to have a garbage patch. One is located in the Indian Ocean, two are in the Atlantic Ocean, and two are in the Pacific Ocean. The most famous of them all is the Great Pacific Garbage Patch, which is located in the North Pacific Gyre between Hawaii and California. In 2018 scientists determined that this garbage patch was roughly twice the size of Texas.

Many types of plastic debris can be sucked into garbage patches, and among the most common are fishing nets. Scientists who have studied the Great Pacific Garbage Patch estimate that these nets make up nearly half of the massive patch. Also common in garbage patches are microplastics, which are tiny bits of plastic that are smaller han 3/16 of an inch (5 mm) in diameter. These plastic bits are the remnants of wind, sunlight, and wave action breaking plastic waste down into smaller and smaller pieces. Researchers who travel to garbage patches to study them may not see the microplastics at first—but when they get closer, the view can be shocking. When ecologist Chelsea Rochman was a graduate student, she was fortunate enough to land a spot on a research vessel that was headed out to the Great Pacific Garbage Patch. Their assignment was to count the plastic waste as it drifted by— but they found that counting was impossible. Rochman explains, "We're looking and it's, like, basically a soup of confetti, of tiny little plastic bits everywhere. Everyone just stops counting. [They] sat there, their backs up against the wall and said, 'OK, this is a real issue, [and it's] not an island of trash you can pick up.'"[26]

Rochman was so concerned by what she observed that she decided to devote her entire career to studying microplastics.

One of the most troubling aspects of these plastic bits is how they can work their way into the food chain. According to Rochman, understanding how microplastics get into fish is important not just because of potential harm to the fish, but because of what it could mean to humans. "We eat fish that eat plastic," she says, adding that there are many unanswered questions. Research on microplastics is a relatively new field of study, and a great deal is unknown. Some of the questions she and others hope to answer, says Rochman, include, "What are all the sources where [the plastic is] coming from, so that we can think about where to turn it off? And once it gets in the ocean, where does it go? Which is super-important because then we can understand how it impacts wildlife and humans."[27]

An Alarming Situation

The issue of plastic pollution is no small or insignificant matter. Plastic in its innumerable forms is used by people all over the world. Most plastic products end up in landfills, as litter, or in the ocean. The environment suffers because of plastic waste, and wildlife of all types are severely threatened. Scientists and environmental activists warn that the problem will only grow worse in the coming years if plastic consumption is not drastically cut. Without tangible solutions, the environmental damage could be irreparable.

Chapter Three

Growing Heaps of E-Waste

From smartphones and smart homes to Fitbits and e-book readers, technology has become more and more integrated into every aspect of people's lives. The latest, coolest tech gadgets practically fly off store shelves and sell out online as soon as they become available. Of course, manufacturers are happy to oblige this massive worldwide demand, introducing new feature-packed electronics continuously—which means tech gadgets become obsolete soon after they are purchased. As a result, a huge number of no-longer-wanted electronics are thrown away every day of every month of every year, even if they are still in working order. This has led to a burgeoning amount of what is known as e-waste, which global affairs correspondent Peter Ford describes in a July 2018 *Christian Science Monitor* article:

> A look around a reporter's desk turns up eight items that will end up as e-waste one day: a printer, keyboard, computer screen, laptop, mouse, and phone charger, as well as mobile and landline phones. And a swivel of the desk chair brings into view a digital camera, television, and cable TV box. E-waste is an informal name for what is also

called waste electrical and electronic equipment. . . . It encompasses any household or office item at the end of its useful life that has circuitry inside, or electrical components drawing on a battery or power supply.[28]

An Astronomical Amount

Ford's reference to "any household or office item" is noteworthy because e-waste consists of more than just electronic gadgetry. In a 2019 report by the World Economic Forum (WEF), an international organization that works to shape global, regional, and industry policies, the authors refer to e-waste as "anything with a plug, electric cord or battery."[29] So along with electronic devices such as laptop and desktop computers, keyboards, printers, smartphones, digital cameras, and televisions, e-waste also consists of household appliances of all shapes and sizes, from dishwashers to vacuum cleaners to microwave ovens to cappuccino brewers. Also contributing to the world's e-waste is heating and cooling equipment, such as furnaces, heaters, and air conditioners. Outdated VHS and DVD players also become e-waste, as do old stereo components, flip phones, and video game consoles.

The amount of e-waste being generated is soaring; in fact, it is the world's fastest-growing type of waste. The WEF estimates that at least 55 million tons (50 million metric tons) of e-waste are produced each year, which is at least double the amount of a decade ago. To make that amount more understandable, the WEF report authors write, "Imagine the mass of 125,000 jumbo jets—it would take London's Heathrow Airport up to six months to clear that many aircraft from the runways. If you find that difficult to envisage, then try the mass of 4,500 Eiffel Towers, jam them all in one space, side by side, and they would cover an area the size of Manhattan."[30]

E-waste is the world's fastest-growing type of waste. In addition to computers and cell phones, e-waste also consists of televisions, old stereo components, and other household equipment.

Countries throughout the world generate e-waste, but some produce far more than others. According to the December 2017 *Global E-Waste Monitor*, a report that identifies which regions and countries generate the most e-waste, Asia is the world's largest e-waste producer. During 2016 Asia's 49 countries collectively produced 20 tons (18.2 metric tons) of e-waste. Europe, with 40 countries, is the world's second-largest e-waste producer, at 13.6 tons (12.3 metric tons) generated in 2016. In third place are the Americas, whose 35 countries generated 12.5 tons (11.3 metric tons) of e-waste. Africa is fourth, with 53 countries that collectively produced 2.4 tons (2.2 metric tons) of e-waste. Oceania, a region that includes Australia, New Zealand, New Guinea, and most of the islands in the Pacific Ocean, is the world's fifth-largest e-waste producer, with 1,600 pounds (726 kg) of e-waste generated during 2016.

E-Waste Contamination in Thailand

The Southeast Asian country of Thailand is known for its tropical climate, pristine beaches, and ornate temples. In recent years Thailand has also become known as a dumping ground for e-waste from the United States, China, Australia, and other industrialized countries. One of the provinces that is now saturated with mounds of e-waste is the agricultural district of Chachoengsao, which is east of Thailand's capital city of Bangkok. A local villager, Payao Charoonwong, has lost her main source of water as a result of e-waste contamination. "I have been using water in this well for 20 years for cooking, boiling and drinking," she says. "But the condition of the water in the well is not usable anymore."

Charoonwong's property sits in the middle of cassava fields, but it was transformed in late 2017 when e-waste began arriving by the truckload. "When it was raining, the water went through the pile of waste and passed our house and went into the soil and water system," says Charoonwong. "The water started to change from clear water to orange colour. There was a bad smell—very bad—and there are toxic chemicals." Water tests conducted by an environmental group and the local government found toxic levels of iron, manganese, lead, nickel, and in some cases arsenic and cadmium—all chemicals that e-waste is known to contain. According to Charoonwong, her once bountiful cassava harvests are now rotting in the fields.

Quoted in Kathryn Diss, "Recycled Electronics Are Turning Thailand into a 'Dumping Ground for Hazardous Waste,'" ABC News, July 16, 2019. www.abc.net.au.

E-Waste = Toxic Waste

All the e-waste that the world's population generates each year represents a tremendous amount of hazardous waste. Although smartphones, televisions, laptops, video game consoles, and other items that make up e-waste do not in any way appear dangerous, the problem is what they contain inside. Many of these devices contain toxic metals such as lithium, mercury, cadmium, and lead, as well as other harmful substances. "E-waste contains a lot of poisonous chemicals," says Ford. "There is mercury in liquid-crystal display screens, lead in cathode-ray tubes, cadmium ion semiconductors and batteries, and ozone layer-destroying chlorofluorocarbons in old refrigerators."[31] Toxic metals such as lead and mercury are carcinogenic, meaning they can lead to cancer in humans and animals. And they have other negative effects on human health as well. Being exposed to heavy metals in the environment can potentially impair the lungs, brain, kidneys, heart, liver, and even the nervous and reproductive systems.

According to the United Nations Environment Programme (UNEP), less than 20 percent of the world's e-waste is properly dealt with, meaning collected and delivered to legitimate recycling operations. Exactly what happens to the rest is largely unknown because few countries compile statistics about e-waste. In richer countries, including the United States, most of it probably ends up in landfills, which can be extremely hazardous to the environment. When electronics are broken or crushed by heavy equipment, heavy metals such as mercury, lead, and cadmium can leak out. They mix with leachate, which then seeps down through the layers of trash. Leachate is considered a toxic substance, and when it contains heavy metals it is especially dangerous. When it gets into soil and groundwater, humans can be exposed to lead and other toxins that can harm human health. According to the WEF, about 70 percent of toxic metals found in landfills have come from the disposal of electronic equipment.

Instead of landfill disposal, some countries burn e-waste. But that is a very unsafe and risky practice that can pollute the air and soil and harm human health. Whenever e-waste is burned, toxic fumes poison the air and make it dangerous for people to breathe. Burning laptops, smartphones, and other electronics with plastic shells, for instance, can release deadly substances known as dioxins, which have been associated with many serious diseases and health conditions. The World Health Organization writes, "Dioxins are highly toxic and can cause reproductive and developmental problems, damage the immune system, interfere with hormones and also cause cancer."[32]

Along with being dangerous to people's health, toxic chemicals that are released when e-waste is burned can damage the upper atmosphere and contribute to ozone depletion. The ozone layer acts as an invisible shield that filters out harmful ultraviolet radiation (UV-B) from the sun. According to the UNEP, "Long-term exposure to high levels of UV-B can severely damage most animals [including humans], plants and microbes, so the ozone layer protects all life on Earth."[33] When the ozone layer becomes thinner, humans have a higher risk of getting sunburn, premature aging of the skin, and skin cancer.

A Toxic E-Waste Dump

Countries that burn e-waste are typically poor developing nations where people are desperate to make a living. One of these countries is Ghana, in West Africa. At the Agbogbloshie dump, which is located in Ghana's capital city of Accra on the banks of a lagoon, workers burn insulated plastic wires to get to the copper inside. People of all ages work at Agbogbloshie, including young children and adolescents. Copper is a valuable metal, and once it has been separated from the insulation, workers can sell it. Although there are much safer methods of separating wires, burning is by far the fastest and cheapest method. "The task of the recycler is to separate the two substances as quickly and economically as possible,"[34] says journalist and author Adam Minter, who traveled to Ghana and visited with workers at Agbogbloshie. Despite the fact that burning mounds of trash fill the air with clouds of billowing, noxious, black smoke, workers continue to burn the wires in order to make a living the only way they know how.

One of the workers at Agbogbloshie is Abdullah Boubacar, who moved to Accra from northern Ghana in 2008. He works at the dump trying to earn a living from e-waste scrap he is able to recover. Like other workers, Boubacar has many serious health problems. "When I was a small boy, I used to be a footballer—but not anymore," he says. "I have stomach ulcers and I run out of en-

At the Agbogbloshie dump in Ghana workers burn insulated plastic wires to get to the copper inside. These workers are subjected to dangerous toxic fumes that are emitted when the e-waste is burned.

ergy very easily."[35] Other health issues that are common among the people who work at Agbogbloshie include burns, back problems, and infected wounds. The toxic air pollution is directly connected to respiratory problems, chronic nausea, and severe headaches.

Because Agbogbloshie is such a polluted, hazardous place for people to work, it has received a great deal of media coverage. In many news stories it has been presented as a dumping ground for all the e-waste produced by much richer industrialized countries, including the United States. In fact, the claim is often made that wealthy countries send at least 20 percent of their e-waste to developing countries. But according to the UNEP, that is not the case in Ghana. Rather, say UNEP officials, some 85 percent of the e-waste in Ghana (including in Agbogbloshie) and other parts of West Africa was produced by people in those regions. "Agbogbloshie is not a global dumping ground," says Minter. "Like most places on Earth, it's struggling to deal with what it generates on its own."[36]

Wasted Resources

People in developing countries are aware that e-waste is rich in valuable materials. Even at the risk of their health, they do everything ossible to remove those materials and sell them. Although the amount of precious metal in each device they dissemble may be small, the total amount quickly adds up. But in the United States and many other industrialized countries, most e-waste ends up being thrown in the trash bin. So instead of precious minerals such as gold, silver, platinum, copper, palladium, and others being recovered in a process known as "urban mining," they are buried under tons of trash and garbage in landfills. According to the WEF, there is one hundred times more gold in one ton of smartphones than there is in one ton of gold ore. During 2016 alone, 480,000 tons (435,000 metric tons) of cell phones were discarded—and in the process, billions of dollars' worth of materials were lost. "Right now, electronics companies spend a fortune buying and processing precious minerals, only to see them buried in landfills,"[37] says Peter Holgate, founder and CEO of the high-tech e-waste recycling firm Ronin8 Technologies.

According to the EPA, one ton of circuit boards contain as much as eight hundred times the amount of gold and twenty to

There is one hundred times more gold in one ton of smartphones than there is in one ton of gold ore. Yet in industrialized countries, most cell phones end up being thrown in the trash.

A Recycling Robot Named Daisy

The technology giant Apple is well known for being an environmentally focused company. One way Apple shows this commitment is by taking back and recycling every product it sells, from iPhones to iMacs. In keeping with its environmental commitment, Apple introduced a new product just before Earth Day in 2018—but it was not a new iPhone or iAnything. It was Daisy, a robot designed to tackle e-waste unlike any robot had before. Technology journalist Michael Moorhead says he was flabbergasted while watching the robot at work. "I got the chance to see Daisy in action as it literally consumed iPhones in one end and then handed back components for recycling on the other," he says. Moorhead has toured plants throughout the world and has seen numerous robots in action. "But I had never witnessed anything like Daisy," he says. Recycling robots typically smash components, which often causes the pieces to comingle and become impure. Daisy disassembles phones, takes out the parts, and puts them into separate buckets, so components can be recycled.

Daisy can handle up to 200 iPhones per hour, which totals more than 1 million devices per year. For every 10,000 iPhones that pass through Daisy's robotic arms, it can reclaim 4,188 pounds (1,900 kg) of aluminum, 1,698 pounds (770 kg) of cobalt, and 1,565 pounds (710 kg) of copper. Once Daisy has recovered these materials, they are recycled back into Apple's manufacturing process.

Michael Moorhead, "Apple's New iPhone Recycling Robot 'Daisy' Is Impressive, and in Austin," *Forbes*, April 19, 2018. www.forbes.com.

forty times the amount of copper that is mined from one ton of ore in the United States. The agency adds that in 2016, the estimated value of recoverable e-waste worldwide was $64.6 billion—but precious metals were only recovered in 20 percent of that e-waste. "Much of the rest is dumped in landfills where toxic chemicals can leach from the e-waste and end up contaminating the water supply,"[38] says the EPA.

In 2018 researchers studying an industrial region in Delhi, India, discovered that an alarming amount of toxic heavy metals had leached into soil and groundwater. In this region, known as the Krishna Vihar industrial area, e-waste is freely dumped on open

...fe recycling methods, including improper dismantling ...ing of e-waste, also take place there. These practices ...c substances to leak onto the ground. In the same area ...ning of e-waste pollutes the air and water with poisonous substances. The researchers studied soil samples and found the average concentration of heavy metals to be as much as thirty times higher than normal. They also found dangerously high levels of heavy metals in groundwater. In an article about the study, coauthors Rashmi Makkar Panwar and Sirajuddin Ahmed discuss the environmental risks associated with this pollution: "Water in these areas is not suitable for drinking. . . . The heavy metals are cause for possible risk to health by direct exposure and consumption of water."[39]

A Looming, Growing Crisis

People throughout the world have come to rely on their electronic devices for daily life. The gadgets are used for texting friends, posting on Instagram and Facebook, tweeting, doing internet research, blogging, shopping online, watching YouTube, reading the latest news, watching movies, and mapping routes for road trips. These and other tech devices eventually become e-waste, as do old televisions, microwaves, and countless other products that make life more convenient. As the global population continues to grow and newer, fancier products become available, the world's e-waste also grows—and grows, and grows, and grows. According to the WEF, if e-waste production continues at its current pace, the volume could exceed 132 million tons (120 million metric tons) by 2050. That could very well create an environmental disaster that would be impossible to repair. Jean-Pierre Schweitzer, an officer with the European Environmental Bureau, shares his thoughts: "E-waste is the next big environmental challenge in today's digital society, a time bomb waiting to explode."[40]

Chapter Four

A Staggering Amount of Wasted Food

"America does not eat 40 percent of its food." That eye-opening declaration is the first sentence of an August 2017 report by the Natural Resources Defense Council (NRDC). The subject of the report is food waste. The report authors, NRDC scientist Dana Gunders and journalist/food waste expert Jonathan Bloom, convey the magnitude of America's food waste with a metaphor: "If the United States went grocery shopping, we would leave the store with five bags and drop two in the parking lot. And leave them there. Seems crazy, but we do it every day."[41]

The immense amount of food that goes to waste is especially troubling in light of the widespread prevalence of hunger. According to the US Department of Agriculture (USDA), more than 37 million Americans, including 6 million children, faced food insecurity in 2018. People who experience food insecurity lack access to nutritious food that is both affordable and available in sufficient quantity. The global view of this problem is especially bleak, with 2 billion people throughout the world facing food insecurity and nearly half of those suffering from severe to moderate hunger. At the same time, an estimated one-third of all food produced worldwide is either lost (never makes it past the farm or food production stage) or wasted. Global hunger coupled with the potential environmental harm caused by wasted food have made food waste one of the most critical issues of the twenty-first century.

From Food Waste to Methane Gas

Research has shown that when food is thrown away, it almost always ends up in landfills. In fact, according to the EPA, about 95 percent of discarded food goes to landfills. The agency also notes that of all the contributors to solid waste in the United States, food waste is the largest component, accounting for 21 percent of all waste dumped into landfills. Unlike plastic, glass, and other nonorganic waste, food waste decomposes. But in a landfill it is buried beneath heavy layers of soil and more waste, so oxygen cannot reach it. As a result, food waste decomposes without the presence of oxygen in what is known as an anaerobic environment. This environment is ideal for producing heat-trapping methane gas.

The Food and Agriculture Organization reports that close to one-third of the world's food is wasted, making food waste a significant contributor to greenhouse gas emissions. And according to the EPA, methane is at least 25 times more potent than CO_2 as a greenhouse gas. "If food waste were a country, it would come in third after the United States and China in terms of impact on global warming,"[42] says Chad Frischmann, vice president and research director of the climate change research organization Project Drawdown.

Food Waste and Leachate Formation

In addition to creating methane gas, food waste that ends up in landfills often contributes to the formation of leachate. This occurs because rotting food contains a great deal of moisture. The combination of naturally occurring moisture and precipitation results in ideal conditions for leachate to form. If the landfill leaks, the leachate can seep into and pollute groundwater. Or if the leachate is treated along with other sewage and then released into nearby lakes and rivers, it could possibly pollute those waterways.

The people of Coventry, Vermont, are worried about this latter scenario. The town's landfill generates more than 9 million gallons (34 million L) of leachate every year. Samples of the leachate have been found to contain a mix of toxic chemicals that have been linked to a variety of health problems. Attorney Elena Mihaly, who has worked with Coventry residents in their fight against the landfill, says these chemicals (some of which have been identified as possible carcinogens) never completely break down. This is worrisome, she says, because the Coventry landfill captures its leachate but has no system for removing the toxic chemicals that are found in leachate. So the substance is

An estimated one-third of all food produced worldwide is either lost at the farming or production stage or is discarded by consumers. About 95 percent of discarded food goes to landfills.

trucked to a wastewater treatment facility, where it undergoes the same treatment process as sewage. After treatment, the sewage and leachate are released back into local waters. According to Mihaly, "That means what is left over after treatment is still dangerous to people and the environment when it's dumped back into our waterways," she says. "Even minuscule amounts of these chemicals are dangerous—and they stick around for a long, long time."[43]

Wasted Resources

Along with environmental contamination, discarded food wastes natural resources. A vast amount of resources is needed to produce food, especially land and water. According to the NRDC, agriculture consumes nearly half of America's land. It also accounts for more than 60 percent of freshwater—tens of millions of gallons that are needed for irrigation. If food is thrown out rather than being eaten, all these resources have also gone to waste. Environmental author John Hawthorne writes:

> Fruit and vegetables are among the most water-laden food products, simply because they contain more water. (For example, one bag of apples is about 81% water!) But meat products are the heaviest water users, simply because the animals drink a lot of water—and more importantly, because so much water is needed for the grain that becomes their feed! It takes about 8 to 10 times more water to produce meat than grain.[44]

Another casualty of wasted food is wasted oil. In order to grow, transport, store, and cook food, massive amounts of fuels such as oil and diesel are required. Fuel is needed for heavy equipment that does the harvesting, trucks that transport food from the farm to the warehouse and grocery store, and other machinery that is needed to sort, clean, and package food to prepare it for sale. "To waste millions of tons (in America) or billions (worldwide) each

Wasted Food in School Cafeterias

Food waste is a serious problem in school cafeterias throughout the United States. In 2019 Lisa Young, a teacher at H.G. Hill Middle School in Nashville, Tennessee, and her students decided to investigate the extent of the problem at their school. They conducted an audit of food waste in the school cafeteria. On the day of the audit, all students left their lunch trays at the audit station. There, student auditors sorted through the food waste. They found more than 168 pounds (76.2 kg) of uneaten food, which included unopened packaged food, as well as uneaten fruits, vegetables, and other foods. In addition, 43 pounds (19.5 kg) of milk was thrown out.

Food policy researcher Melissa Terry conducted a similar audit of school cafeteria food waste in the Atlanta, Georgia, area in 2015. She measured 13 gallons (49 L) of wasted milk in one day. Terry points out that this waste happens in schools all across the country. "That's significant when you think about the carbon and methane footprint of dairy, the costs of refrigeration, transportation, and packaging," she says.

Most schools have a policy that unopened or untouched food cannot be served again or removed from the cafeteria but instead must be tossed. And the US Department of Agriculture requirement that milk be available to students at school lunches often means that kids grab a carton of milk that they only partly drink—or never even open.

Quoted in Lela Nargi, "Millions of Dollars' Worth of Food Ends Up in School Trash Cans Every Day. What Can We Do?," Ensia, April 9, 2019. https://ensia.com.

year also means that all of the oil and fuel that has gone into the production of said food is wasted,"[45] says Hawthorne. Also, any machinery that uses fuel can release harmful toxins into the environment, including greenhouse gases. When food goes to waste, that pollution occurred needlessly.

The Biggest Food Wasters

Although countries throughout the world waste food, the United States is second only to Australia in being the most wasteful. Research has shown that the American population wastes 125 billion to 160 billion pounds (57 billion to 73 billion kg) of food each year. And this food waste comes from multiple sources, as Move for Hunger's Zachary Sobol writes: "Essentially, food is being wasted at every level of the supply chain. From the farm, to distribution, to the grocery store, and then finally at home, food waste is an inherent problem."[46] In referring to the farm, Sobol is talking about food that becomes waste because it never makes it to market. This happens for a variety of reasons. According to the USDA, one reason is that market prices at a given time may be too low for farmers to justify the cost of harvesting crops. Or a farmer may not be able to find labor to do the harvesting, so crops are left to rot in fields.

For its 2017 report, the NRDC studied which foods most often go to waste. Fruits and vegetables are at the top of the list, collectively totaling about one-third of all food waste. Dairy products represent nearly 20 percent of food waste, followed by grain products (such as cereals, bread, and rice), meat, poultry, and

Agriculture consumes nearly half of America's land. It also accounts for more than 60 percent of freshwater—tens of millions of gallons that are needed for irrigation. If food is thrown out rather than being eaten, all these resources have also gone to waste.

fish. The biggest culprits in all of this waste are private house-holds, which collectively generate 43 percent of the country's to-tal food waste.

Among households that regularly waste food, one of the most significant factors is affluence. Americans who can afford all the food they need and want are much more apt to waste food than those who worry about where their next meal will come from. Oth-er factors involved in wasting food include poor (or lack of) meal planning, buying and/or preparing too much food, not properly storing leftovers, and forgetting about food that has been placed in the refrigera-tor. As journalist Sarah Taber writes in the *Washington Post*: "Personally, I find that once produce goes in my crisper drawer, it might as well be in a black hole."[47]

People also waste food because they are confused by terms that are printed on bottles, jars, and food packaging. The terms "best if used by," "sell by," and other types of date labels are often misinter-preted by consumers to mean "do not use after [date]," which is not necessarily true. "The labels, you see, don't mean what they appear to mean," says NPR food and agriculture correspondent Dan Charles. "Foods don't 'expire.' Most foods are safe to eat even after that 'sell by' date has passed. They just may not taste as good, because they're not as fresh anymore. . . . But those dates often have the perverse effect of convincing over-cautious consumers to throw perfectly good food into the trash."[48]

Wasteful Grocery Stores

Although households are the main contributor to America's wasted food, they are by no means the only contributor. Gro-cery stores and distributors represent 13 percent of wasted

IMPACT FACTS

Each year, more than 52 million tons (47.2 million metric tons) of food is sent to landfills, and an additional 10.1 million tons (9.2 million metric tons) remains unharvested on farms.

— ReFED, a nonprofit dedicated to reducing food waste

food—as much as 21 billion pounds (9.5 billion kg) per year. Jennifer Molidor, who directs the Center for Biological Diversity's sustainable food initiatives, writes, "The blame shouldn't be entirely in the consumers' basket—where food is sold plays a big role in what consumers buy and even their likelihood of throwing it out too soon."[49]

One of the points Molidor makes about grocery stores is the extraordinary amount of produce that is wasted because it has a few flaws or imperfections. Although this does not in any way affect taste, quality, or nutritional value, most customers will not buy the produce because of how it looks. According to Molidor, grocery stores share the blame for that with consumers. One reason is that stores do not typically display fruits and vegetables that are weird shapes or do not look perfect. Thus, consumers have the perspective that produce is only worth buying if it looks a certain way. "When supermarkets only stock perfectly round apples and straight carrots," she says, "customers get used to thinking produce has to look a certain way. Meanwhile, by relaxing cosmetic standards, more than 250,000 tons of edible produce could be saved annually from being left to rot in the field."[50]

Dan Christmann, who is with the West Michigan Environmental Action Council, worked for a grocery store and saw for himself how much good food ended up in the garbage. One reason stores throw out such enormous amounts of food is the sheer volume of products they stock. They want customers to always find whatever they want, as Christmann explains: "The common wisdom is that customers are used to having every product available, from fresh Jackfruit to sea scallops farmed off the South African coast, so managers stock these rare products even if most go to waste." Even when grocery store

IMPACT FACTS

Growing and transporting the amount of food that goes to waste in the United States emits as much carbon pollution as 39 million passenger vehicles.

— Ben Simon, cofounder and CEO of Imperfect Produce

A large amount of produce is wasted because of a few flaws or imperfections. Although the flaws do not affect taste or nutritional value, many people will not buy produce that looks odd.

management knows that a huge percentage of food will be wasted, says Christmann, "that's the price of doing business."[51]

Exorbitant Restaurant Waste

Restaurants waste even more food than grocery stores and supermarkets. NRDC research reveals that 18 percent of America's food waste comes from food thrown out by restaurants. This includes all kinds of restaurants, from fast-food places to fine-dining establishments. According to a 2018 study of restaurateurs in Berkeley, California, the most wasteful of these are dining places such as school and college cafeterias and food establishments that offer all-you-can-eat buffet-style meals. The researchers found a number of reasons why restaurants waste so much food. These include inadequate food storage, improper handling of food, customer portions that are excessively large (leaving leftovers on plates), the challenge of accurately forecasting the number of patrons, forgotten and spoiled food, and low awareness of the economic and environmental costs associated with wasted food.

People often throw out a lot more food than they think they do. This was revealed during a 2019 Ohio State University study, in which researchers examined food in refrigerators because that is where most perishable food is found in the typical household. "We wanted to understand how people are using the refrigerator and if it is a destination where half-eaten food goes to die," says Brian Roe, the study's senior author.

The study used data from the State of the American Refrigerator survey, which examined refrigerator contents and practices using 307 survey participants. Researchers asked participants how much fruits, vegetables, meats, and dairy they had in the fridge, as well as how much they expected to eat. Follow-up surveys were conducted a week later to determine how much food was actually consumed and how much was discarded. The study found that participants expected to eat 97 percent of the meat but actually only ate about half of it. Similarly, they expected to eat 94 percent of their vegetables but only ate 44 percent. They consumed 42 percent of the dairy in the fridge, after thinking they would consume 84 percent. And they expected to eat 71 percent of the fruit but finished only 40 percent. "People eat a lot less of their refrigerated food than they expect to, and they're likely throwing out perfectly good food," says Roe.

Quoted in Ohio State University, "Food-Waste Study Reveals Much Fridge Food 'Goes There to Die,'" SciTechDaily, August 28, 2019. https://scitechdaily.com.

Excessively large portions, which are common in restaurants all over the United States, lead to widespread food waste. Research has shown that food portions have increased by more than 130 percent since the 1970s. In March 2019 *Washington Post* food critic Tom Sietsema wrote an article criticizing restaurant portion size. Because of the nature of his work, Sietsema dines out at least ten times a week. He has been surprised by the huge amounts of food that restaurants heap on plates. One restaurant, says Sietsema, offered homemade ice cream sandwiches that weighed 15 pounds (6.8 kg) and were made from five flavors of ice cream "and cookies the size of hubcaps."[52] Another restaurant's offering was made with ribs, bacon, sausage, a cheese sauce, and jalapeño peppers, and its name was indicative of the huge size: the Meat Tornado. The title of Sietsema's article aptly sums up the action he wishes eating establishments would take: "In an Era of Excessive Food Waste, a Plea to Restaurants: Cut Down Your Portions."

Atrocious Waste

Food waste is a serious problem for which there are no easy answers— but answers are sorely needed. Since the 1970s the problem has grown steadily worse, with an estimated 40 percent of all food now going to waste in the United States.

IMPACT FACTS

Methane is at least twenty-five times more potent than CO2 as a greenhouse gas.

— US Environmental Protection Agency

Dana Gunders says that solving this problem is indeed challenging but not impossible. It begins, she says, with public awareness, which she believes is slowly growing. "But I think there's still a disconnect between being aware that this is a global problem and connecting that to what you're actually doing when you scrape your plate into the garbage."[53]

Chapter Five

Global Efforts and Challenges

It is widely known that the world has an extraordinary waste and garbage problem—and because of that, the world is also facing environmental threats. In a September 2018 publication, World Bank official Sameh Wahba spoke candidly about these threats, stating, "Poorly managed waste is contaminating the world's oceans, clogging drains and causing flooding, transmitting diseases, increasing respiratory problems from burning, harming animals that consume waste unknowingly, and affecting economic development."[54] Wahba and other environmental experts warn that major steps must be taken in order to protect the world's water, air, and land from the potentially disastrous effects of burgeoning waste. Although this will be an enormous challenge, experts say, it is essential.

Attacking the Plastic Problem

By far, one of the most formidable challenges is plastic waste. Plastic is used worldwide for so many different purposes that reducing plastic waste can seem like an impossible task. Environmentalists emphasize that the key to reducing plastic waste is to use a lot less plastic. "We use an incredible quantity of single-use plastic items, such as straws, plastic bags, packaging, plastic cups, plates and cutlery," says Nina Jensen, who is CEO of a nonprofit foundation called REV Ocean. "We must put an end to it."[55]

Many countries have taken steps in this direction. According to a 2018 United Nations report, more than sixty countries have enacted bans or taxes aimed at reducing single-use plastics. Among these countries is Canada, which has the world's longest coastline and a fourth of the world's freshwater. Prime minister Justin Trudeau announced in 2019 that his nation would ban single-use plastics by 2021. This includes items such as plastic grocery bags, straws, cutlery, and water bottles.

The European Union (EU), which represents twenty-eight countries, took similar action in 2019 when it voted to ban single-use plastic items such as plastic cutlery, straws, and stirrers by 2021. The EU plan also called for plastic bottles to be made of 25 percent recycled content by 2025, as well as for 90 percent of plastic bottles to be recycled by 2025. Frans Timmermans, the European Commission vice president who spearheaded the EU plan, said, "Today we have taken an important step to reduce littering and plastic pollution in our oceans and seas. We got this, we can do this. Europe is setting new and ambitious standards, paving the way for the rest of the world."[56]

IMPACT FACTS

Norway recycles 97 percent of its plastic bottles.

— Climate Action

Garbage Patch Cleanup Efforts

Scientists and others are also trying to come up with ways to eliminate the garbage patches in the world's oceans. The largest of these is the Great Pacific Garbage Patch. Since its discovery in the early 1970s, researchers have been monitoring its size and growth. In 2012 they discovered that the patch had grown one hundred times larger in forty years. It spread out so far and wide that it covered 6.2 million square miles (16 million sq. km)—an area twice the size of Texas.

Dutch entrepreneur Boyan Slat first learned about plastic pollution in the ocean when he was sixteen years old. When he heard about the Great Pacific Garbage Patch, he was shocked at its magnitude—and even more shocked that there were no serious attempts to clean it up. So even though he was only a teenager, Slat vowed to attack the problem himself. He founded a nonprofit organization called the Ocean Cleanup and began developing and testing his idea for a trash-collecting device to deploy in the ocean. His first prototype was unveiled in 2012. The next seven years were filled with several engineering redesigns and numerous disappointments.

In September 2019 Slat's group had finally worked out all the problems and had a new model ready to send to the vast, swirling Great Pacific Garbage Patch. The trash-collecting device is

Canadian prime minister Justin Trudeau (pictured) announced in 2019 that his nation would ban single-use plastics by 2021. This includes items such as plastic grocery bags, straws, cutlery, and water bottles.

composed of a massive boom that acts as a gigantic "arm" as it moves with the ocean's currents. This arm passively catches plastic debris and then concentrates it into the center. Periodically, a boat comes by to collect the accumulated trash, much like an oceangoing garbage truck. The boat then takes the trash back to shore for recycling.

On October 2, 2019, the Ocean Cleanup system finally worked as Slat and his designers had envisioned. Slat was ecstatic over this triumph, which was obvious in a tweet he wrote that evening: "Our ocean cleanup system is now finally catching plastic, from one-ton ghost nets to tiny microplastics!"[57] Along with the tweet was a photo showing a hodgepodge of plastic trash that had been recovered by his device, including crates, fish nets, pieces of rope, food containers, and myriad other plastic objects. With that successful mission behind him, Slat is optimistic about future cleanup efforts. His long-term goal is for 50 percent of the Great Pacific Garbage Patch to be cleaned up by 2024 and 90 percent by 2040. His plan is to design and deploy an entire fleet of the garbage retrieving systems to clean up all five of the world's known garbage patches.

IMPACT FACTS

All plastic waste could potentially be recycled into new, high-quality plastic.

— Chalmers University of Technology, Gothenburg, Sweden

Recycling Struggles

Plastic in the ocean is one of numerous environmental issues that need to be solved in order to protect the planet. Because disposal of nonorganic items (especially plastic and glass) is a problem on land as well as in the ocean, researchers are constantly exploring ways to address the problem. For decades recycling has been viewed as an effective way to manage waste, and environmental groups still emphasize the importance of it.

The recycling industry is encountering problems, however. One of the biggest of these problems involves China, which has stopped taking and processing the world's recyclables. This action has had ramifications for cities and towns all across the United States, as well as in other countries that counted on China for their recycling. In Broadway, Virginia, residents participated in a city-sponsored recycling program for more than twenty years. In 2019 the company that collected recyclables told city officials there would be a 63 percent increase in cost. The company also stopped offering recycling pickup services. Without this service, many residents began throwing away their recyclables. Similar situations happened in Hannibal, Missouri; Erie, Pennsylvania; Columbia County, New York; and in other cities and counties throughout the United States. As recycling costs have soared, people's commitment to recycling has plummeted.

One reason for the huge cost increases can be found thousands of miles away in China. For years, the United States and many other countries packed tons of recyclables in huge containers, shipped them to China, and were paid good money for them. Then in 2017 Chinese officials announced that their country would no longer accept most recyclables. They said too much dirty waste, and in some cases hazardous waste, was being mixed in with the recyclables they received, and it was causing environmental pollution in China. The policy, which went into effect in 2018, banned unsorted waste paper, nearly all plastics, and numerous other items. Environmental author Edward Humes writes, "Massive amounts of . . . recyclables began piling up at US ports and warehouses. Cities and towns started hiking trash-collection fees or curtailing recycling programs, and headlines asserted the 'death of recycling' and a 'recycling crisis.'"[58]

IMPACT FACTS

As of July 2018, at least 127 nations have taxed or banned single-use plastic bags.

— World Resources Institute

Although it is highly unlikely that the situation with China has caused the "death" of recycling, it has created a formidable challenge. Suddenly, there was a glut of recyclable materials in the

The nonprofit organization Ocean Cleanup developed a trash-collecting device that sits on top of the water catching ocean waste. The group intends to clean up most of the Great Pacific Garbage Patch by 2040.

United States and few buyers for them, which caused recycling companies to recoup their losses by charging communities more to recycle. According to a March 2019 *New York Times* article, some recyclers began charging four times as much as they charged in 2018. The article explains that soaring costs forced communities into making hard choices about whether to raise taxes to pay higher recycling costs, cut municipal services, or stop recycling altogether. "Recycling has been dysfunctional for a long time," says Mitch Hedlund, executive director of Recycle Across America. "But not many people really noticed when China was our dumping ground."[59]

High-Tech Plastic Recycling

An exciting development in plastic recycling was announced in October 2019 by Chalmers University of Technology in Gothenburg, Sweden. A team of researchers developed an efficient process for breaking down plastic waste to a molecular level, in

The Ecoffee Cup

According to the Earth Day Network, an enormous number of disposable coffee cups are used globally each year—more than 16 billion. The cups are typically made of paper that is coated with plastic, and this combination of materials makes them almost impossible to recycle. In 2014 David McLagan, a giftware manufacturer, decided to do something about the huge waste problem created by disposable cups. "Most single-use items get used for a few minutes and then get discarded," he says. "I just looked at the mountains of waste and decided enough was enough." So he created a reusable, permanent cup named the Ecoffee Cup as an alternative to single-use, disposable coffee cups.

The Ecoffee Cup resembles a disposable coffee cup but is made of natural bamboo fiber, cornstarch, and resin. The lid and sleeve are made with latex-free, food-grade silicone. The entire cup, including the lid and sleeve, is dishwasher safe. The Ecoffee Cup is so popular—more than 4 million were sold in 2018—that the company has occasionally run out of stock.

McLagan's company is based in Amsterdam, with plans to expand to the United States. Ultimately, his hope is to get people to shift away from using disposable products in favor of reusable products. He says he wants "people to stop and think about the impact they're having, and how small, simple changes can have big impacts." He adds, "I genuinely feel we're only getting started in our quest to change the world."

Quoted in Lucy Sherriff, "Saving the Earth, One Cup of Coffee at a Time," *Forbes*, January 10, 2019. www.forbes.com.

the form of a gas. The resulting gases can then be transformed back into new plastics that are of the same quality as the original plastic.

According to the project leader, energy technology professor Henrik Thunman, plastic's resilience—the very quality that causes it to be an environmental nightmare—is actually an asset. "We should not forget that plastic is a fantastic material—it gives us products that we could otherwise only dream of," says Thunman. "The problem is that it is manufactured at such low cost, that it has been cheaper to produce new plastics from oil and fossil gas than from reusing plastic waste."[60] Thunman adds that with the process he and his fellow researchers developed, plastic is recycled at the molecular level. It can then be used to make new plastic materials of a quality that is identical to the original plastic.

What Can Be Done About E-Waste?

Along with plastic, e-waste is one of the worst waste-related problems worldwide—and the fastest-growing type of waste. Although e-waste recycling has garnered a bad reputation because of the way it is often conducted and the people who are harmed, experts say there is a great deal of untapped potential. It is widely known, for instance, that e-waste is rich in precious metals like gold, silver, platinum, and copper. According to Peter Holgate, founder and CEO of the high-tech e-waste recycling firm Ronin8 Technologies, methods of extracting the metals from e-waste exist, but they are inefficient and costly. But new companies, says Holgate, are "attacking the problem from multiple angles."[61]

Holgate's own team has developed a process to recover precious metals in iPhones using powerful underwater sound waves. An expert in e-waste recycling, Holgate envisions a time when what is now considered waste will become a valuable resource. For instance, extracting metals from e-waste reduces the need for virgin metals such as gold, copper, and aluminum. These metals are contained within rock known as ore that is buried deep within the earth. In order to extract the metals, the land must be mined and the ore dug out, which causes serious environmental damage. When less land is needed for mining, the environment benefits immensely.

Another vision experts have for the future is a massive reduction in e-waste being produced in the first place. This could come about with better design of electronic products—which environmental groups and waste management experts say is sorely needed because these products are designed to be obsolete within one or two years.

...g that electronic products be built to last is the focus ...pair, a movement that was started by the activist ...n Environmental Bureau (EEB). The organization's ...explains that manufacturers of these products intentionally make it difficult and expensive to repair electronic gadgets and replace parts such as cracked screens or weak batteries. "By reducing the lifespan of a product," says the EEB, "they may drive sales, but this comes at the expense of citizens and the planet."[62] The EEB states that the costly expense of repairing many electronics has led to a "throwaway culture" in which people would rather just trash their tech gadgets rather than bother to have them fixed. This needs to change, and helping make that happen is a priority of the EEB and other environmental groups.

Their efforts have started to pay off; at least eighteen US states have introduced Right to Repair legislation. These laws would require electronics manufacturers to service their products by linking customers to independent

Experts have suggested the costly expense of repairing electronics has led to a "throwaway culture" in which people would rather throw their tech gadgets away than get them fixed.

repair shops with diagnostic equipment and replacement parts. Soon after states began passing the legislation, support for the Right to Repair movement began to grow. Whether this will lead to a reduction in e-waste is unknown because the problem is such a difficult one to solve.

Addressing the Scourge of Wasted Food

Another waste-related problem that is immensely difficult to solve is food waste. Yet experts say that there are numerous ways for people and businesses to help address the problem and cut down on waste. Adam Redling, editor of the publication *Waste Today*, writes:

> It's just about finding the motivation to make some simple changes in our day-to-day routines. Who knows, maybe one day in the near future, the food that is currently being sent to landfills in record quantities will instead be donated to feed the hungry or used to compost our fields, power our homes and cars, and provide better nutrition to our livestock. Talk about an opportunity too good to waste.[63]

One individual who is addressing the food waste problem is Ben Simon, cofounder of a company called Imperfect Produce in San Francisco. He and his business partner are committed to helping reduce the produce that is wasted by farms because it is not "perfect" enough for grocery stores to sell. They have built relationships with farmers across the United States and regularly buy their misshapen, too-small but fresh fruits and vegetables that would otherwise end up in the landfill. "It's fairly consistent that 15 to 35 percent of a grower's crop goes to waste simply because it cannot go to traditional grocery stores,"[64] says Simon. Customers sign up with Imperfect Produce on a subscription basis to have fresh produce delivered to their homes. In addition, the company donates thousands of pounds of produce to food banks every year.

As awareness of food waste has grown, more companies and individuals have searched for ways to be involved in fixing the

Teens Tackle Food Waste

In 2019 two Florida teens became so alarmed about food waste that they decided to take action. The teens, siblings Ugo and Emma Angeletti, first became concerned about the issue while Ugo was working on a high school science project. He measured the trash people produce on a weekly basis and discovered that food waste was the biggest culprit. When he talked to his family about his project, Emma recalls, "We all were very shocked by the numbers . . . and thought we had to do something."

Ugo and Emma started a composting charity named back2earth. A stranger donated a plot of land for a composting station. Ugo and Emma, along with their younger sisters and a team of youth volunteers, do all the composting. Local residents fill buckets with food waste throughout the week. Each Saturday, the youth volunteers collect the buckets, leave a clean bucket, and take the food waste to the composting station.

The teens have documented their progress on the back2earth website. It shows that their efforts have diverted 11,450 pounds (5,194 kg) of food waste from landfills, produced 3,250 pounds (1,474 kg) of compost, and prevented 130,500 pounds (46,947 kg) of CO_2 from being emitted. Their efforts have attracted the attention of other teens, who have emailed Ugo and Emma about starting similar programs in their own communities. "We realize that young people follow young people," says Emma, "so a young person will follow a movement if another young person inspires them to."

Quoted in Katie Kindelan, "These Teen Siblings Have Turned 15,000 Pounds of Food Waste into a Composting Movement," GMA, May 29, 2019. www.goodmorningamerica.com.

problem. In Carson City, Nevada, an environmental group concerned about food waste in restaurants created the green dining district in April 2019. Eight restaurants signed on to be part of the downtown district, one of which is Gather, a farm-to-table eatery. One of the restaurant's green practices is to collect waste products such as cauliflower and cabbage stems, carrot tops and peels, and onion skins and send them to a farm, where they become animal feed. Another green district establishment, Artisan Café, serves sandwiches, salads, and baked goods with the environment in mind. "We repurpose anything and everything," says owner Jeanne Dey. "If we have (kitchen) scraps that can be re-made into soups or things that we can donate that are extras, we do that. There's really nothing that we have that goes to waste."[65] Although these are small-scale efforts, small steps taken by many groups and individuals can collectively lead to major solutions.

Uncertainties Abound

The green dining district in Carson City, Imperfect Produce in San Francisco, the Right to Repair movement, Boyan Slat's Ocean Cleanup endeavor, and scientific research worldwide are all ex-amples of companies and individuals that are working to make a difference. They are doing their part to help fix the world's waste problem. Yet fixing this immense problem—which ranges from plastic waste in the ocean to heaps of toxic e-waste to burgeon-ing amounts of food waste—is a formidable undertaking that will require a global commitment. Can it be done? Even with lots of optimism, creativity, and a mix of possible solutions, that is a question that no one can answer with any certainty.

Source Notes

Introduction: Throwaway Societies

1. Dinyar Godrej, "Modern Life Is Rubbish," *New Internationalist*, November 1, 2018. https://newint.org.
2. Quoted in John Vidal, "Baby Diapers Are Hiding Some Dirty, Dangerous Secrets," Huffington Post, April 18, 2019. www.huffpost.com.
3. World Bank, "What a Waste: An Updated Look into the Future of Solid Waste Management," September 20, 2018. www.worldbank.org.

Chapter One: The Environmental Impact of Landfills

4. Quoted in Gerard Ramalho, "HELPING HEAP: Southern Nevada's High-Tech Waste Disposal Is the Nation's Most Advanced," News3 Las Vegas, November 15, 2018. https://news3lv.com.
5. City of High Point, North Carolina, "Differences Between a Dump and a Landfill." www.highpointnc.gov.
6. Kirstie Pecci, "All Landfills Leak, and Our Health and Environment Pay the Toxic Price," Conservative Law Foundation, July 23, 2018. www.clf.org.
7. Scott C. Christenson and Isabelle M. Cozzarelli, "The Norman Landfill Environmental Research Site: What Happens to the Waste in Landfills?," US Geological Survey, August 2003. https://pubs.usgs.gov.
8. Agency for Toxic Substances and Disease Registry, "ATSDR's Substance Priority List," 2017. www.atsdr.cdc.gov.
9. US Environmental Protection Agency, "Basic Information About Landfill Gas," July 2019. www.epa.gov.
10. New York State Department of Health, "Important Things to Know About Landfill Gas," October 2019. www.health.ny.gov.
11. Quoted in Kristin Musulin, "US Landfill Capacity to Drop 15% over Next 5 Years," Waste Dive, May 8, 2018. www.wastedive.com.

12. Quoted in Matt Davis, "Just How Big Is India's 'Mount Everest of Trash'?," BigThink, June 13, 2019. https://bigthink.com.
13. Quoted in Ann M. Simmons, "The World's Trash Crisis, and Why Many Americans Are Oblivious," *Los Angeles Times*, April 22, 2016. www.latimes.com.

Chapter Two: The Plastic-Plagued Planet

14. Nicholas Rosellini, "Speech by Mr. Nicholas Rosellini at 2018 World Environment Day Event," United Nations in China, June 2, 2018. www.un.org.cn.
15. Laura Parker, "A Whopping 91% of Plastic Isn't Recycled," *National Geographic*, December 20, 2018. www.national geographic.com.
16. Godrej, "Modern Life Is Rubbish."
17. Laura Parker, "We Made Plastic. We Depend on It. Now, We're Drowning in It," *National Geographic*, June 2018. www.national geographic.com.
18. Quoted in ScienceDaily, "More than 8.3 Billion Tons of Plastics Made: Most Has Now Been Discarded," July 19, 2017. www .sciencedaily.com.
19. Earth Policy Institute, "Plastic Bags Fact Sheet," October 2014. www.earth-policy.org.
20. Barry Yeoman, "A Plague of Plastics," National Wildlife Federation, June 1, 2019. www.nwf.org.
21. Yeoman, "A Plague of Plastics."
22. Quoted in Alejandra Borunda, "This Young Whale Died with 88 Pounds of Plastic in Its Stomach," *National Geographic*, March 18, 2019. www.nationalgeographic.com.
23. Quoted in Borunda, "This Young Whale Died with 88 Pounds of Plastic in Its Stomach."
24. Quoted in University of Exeter, "Marine Turtles Dying After Becoming Entangled in Plastic Rubbish," December 12, 2017. www.exeter.ac.uk.
25. National Oceanic and Atmospheric Administration Marine Debris Program, "Garbage Patches," September 26, 2019. https://marinedebris.noaa.gov.

26. Quoted in Christopher Joyce, "Beer, Drinking Water and Fish: Tiny Plastic Is Everywhere," NPR, August 20, 2018. www.npr .org.

27. Quoted in Joyce, "Beer, Drinking Water and Fish."

Chapter Three: Growing Heaps of E-Waste

28. Peter Ford, "'E-Waste': Getting Grip on a Growing Global Problem," *Christian Science Monitor*," July 9, 2018. www.cs monitor.com.

29. Houlin Zhao et al., "A New Circular Vision for Electronics: Time for a Global Reboot," World Economic Forum, January 2019. www3.weforum.org.

30. Zhao et al., "A New Circular Vision for Electronics."

31. Ford, "'E-Waste.'"

32. World Health Organization, "Dioxins and Their Effects on Human Health," October 4, 2016. www.who.int.

33. United Nations Environment Programme, "All About Ozone and the Ozone Layer." https://ozone.unep.org.

34. Adam Minter, "The Burning Truth Behind an E-Waste Dump in Africa," *Smithsonian*, January 13, 2016. www.smithsonian mag.com.

35. Quoted in Peter Yeung, "The Rich World's Electronic Waste, Dumped in Ghana," CityLab, May 29, 2019. www.citylab.com.

36. Minter, "The Burning Truth Behind an E-Waste Dump in Africa."

37. Peter Holgate, "How Do We Tackle the Fastest Growing Waste Stream on the Planet?," World Economic Forum, February 9, 2018. www.weforum.org.

38. Quoted in Renee Cho, "What Can We Do About the Growing E-Waste Problem?," *State of the Planet* (blog), Columbia University Earth Institute, August 27, 2018. https://blogs .ei.columbia.edu.

39. Quoted in Jacob Koshy, "E-Waste Polluting Delhi's Groundwater, Soil: Study," *The Hindu* (Chennai, India), January 12, 2018. www.thehindu.com.

40. Quoted in Citizen Truth staff, "Why E-Waste Is So Dangerous and How the 'Right to Repair' Will Save the Environment," Medium, November 4, 2018. https://medium.com.

Chapter Four: A Staggering Amount of Wasted Food

41. Dana Gunders and Jonathan Bloom, *Wasted: How America Is Losing up to 40 Percent of Its Food from Farm to Fork to Landfill*. Washington, DC: National Resources Defense Council, 2017. www.nrdc.org.

42. Chad Frischmann, "Opinion: The Climate Impact in the Food in the Back of Your Fridge," *Washington Post*, July 31, 2018. www.washingtonpost.com.

43. Elena Mihaly, "Is Toxic Landfill Wastewater Coming to a Stream or Farm Near You?," Conservative Law Foundation, September 9, 2019. www.clf.org.

44. John Hawthorne, "5 Ways Food Waste Is Destroying Our Beautiful Planet," *New Food*, August 9, 2017. www.newfood magazine.com.

45. Hawthorne, "5 Ways Food Waste Is Destroying Our Beautiful Planet."

46. Zachary Sobol, "How Food Waste Is Harming Our Environment," Move for Hunger, July 26, 2018. www.moveforhunger .org.

47. Sarah Taber, "Farms Aren't Tossing Perfectly Good Produce. You Are, *Washington Post*, March 8, 2019. www.washington post.com.

48. Dan Charles, "For Food Manufacturers, 'Sell By' Labels May Have Reached Their Expiration Date," NPR, February 15, 2017. www.npr.org.

49. Jennifer Molidor, "Opinion: When It Comes to Food Waste, Shoppers Aren't the Only Ones to Blame," Food Tank, July 2018. https://foodtank.com.

50. Molidor, "Opinion."

51. Dan Christmann, "Preventing Food Waste in Grocery Stores," West Michigan Environmental Action Council, July 26, 2019. www.wmeac.org.

52. Tom Sietsema, "In an Era of Excessive Food Waste, a Plea to Restaurants: Cut Down Your Portions," *Washington Post*, March 25, 2019. www.washingtonpost.com.

53. Quoted in Rachael Jackson, "Most People Waste More Food than They Think—Here's How to Fix It," *National Geographic*, April 24, 2019. www.nationalgeographic.com.

Chapter Five: Global Efforts and Challenges

54. Quoted in Ede Ijjasz-Vasquez, "What a Waste: An Updated Look into the Future of Solid Waste Management," World Bank, September 20, 2018. www.worldbank.org.

55. Nina Jensen, "8 Steps to Solve the Ocean's Plastic Problem," World Economic Forum, March 2, 2018. www.weforum.org.

56. Quoted in Jennifer Rankin, "European Parliament Votes to Ban Single-Use Plastics," *The Guardian* (Manchester, UK), March 27, 2019. www.theguardian.com.

57. Boyan Slat, "Our ocean cleanup system is now finally catching plastic," Twitter, October 2, 2019. https://twitter.com.

58. Edward Humes, "The US Recycling System Is Garbage," Sierra Club, June 26, 2019. www.sierraclub.org.

59. Quoted in Michael Corkery, "As Costs Skyrocket, More U.S. Cities Stop Recycling," *New York Times*, March 16, 2019. www.nytimes.com.

60. Quoted in SciTechDaily, "All Plastic Waste Could Be Recycled into New, High-Quality Plastic with Advanced New Process," October 20, 2019. https://scitechdaily.com.

61. Holgate, "How Do We Tackle the Fastest Growing Waste Stream on the Planet?"

62. Quoted in Alex Muiruri, "Why E-Waste Is So Dangerous; the 'Right to Repair' May Save the Environment," Citizen Truth, November 4, 2018. https://citizentruth.org.

63. Adam Redling, "The Great American Food Waste Problem," *Waste Today*, April 16, 2018. www.wastetodaymagazine.com.

64. Quoted in Deidre McPhillips, "Q&A: Imperfect Produce's Ben Simon on Ending the Food Beauty Pageant," *U.S. News & World Report*, June 29, 2018. www.usnews.com.

65. Quoted in Kaleb Roedel, "Carson City Restaurants Work to Reduce Food Waste," *Nevada Appeal*, July 22, 2019. www.nevadaappeal.com.

Make a Difference

1. If you bring your lunch from home, choose reusable rather than disposable containers and utensils.

2. Instead of throwing away clothes that are old or do not fit anymore, donate them to a homeless shelter, charity, or thrift store.

3. To help reduce the amount of plastic pollution in the environment, choose paper straws instead of plastic ones.

4. Use both sides of a piece of paper, and recycle it rather than throwing it out when you are finished with it.

5. Instead of using paper towels and paper napkins at home, use cloth towels and napkins.

6. Start your own compost pile or bin in the backyard, using grass clippings, tree leaves, coffee grounds, and fruit and vegetable scraps.

7. Rather than throwing out your old devices like telephones, laptops, or tablets, either resell or donate them to businesses that recycle.

8. To help cut down on plastic pollution, switch from single-use plastic to reusable water bottles.

9. Join other volunteers to help pick up trash at a nearby park or beach.

10. Start a collection drive in your neighborhood or school for recyclable items.

11. Talk to your parents about ways your whole family can reduce waste and garbage.

Earth Day Network (EDN) — www.earthday.org

This nonprofit organization coordinates the annual Earth Day, celebrated April 22 each year in more than 190 countries around the world. The EDN works to protect the environment through activism and advocacy, including its efforts to end plastic pollution. The group's website has information on a variety of ways people can help protect the environment.

National Geographic Kids — https://kids.nationalgeographic.com

This website, which is sponsored by the National Geographic Society, accompanies the magazine of the same name. It includes games, videos, quizzes, tips on how to help reduce trash and waste, and a section titled Kids vs. Plastic that features slideshows, projects, and guides that provide plastic-free strategies for kids.

Natural Resources Defense Council (NRDC) — www.nrdc.org

The NRDC is a nonprofit environmental advocacy group that works to protect the natural systems on which all life on the planet depends. Its website provides news releases and articles related to environmental issues such as preventing pollution and reviving the oceans, as well as information on ways people can get involved in helping protect the planet.

Plastic Pollution Coalition — www.plasticpollutioncoalition.org

This global nonprofit organization works to educate the public regarding the growing rate of plastic pollution worldwide and seeks to find ways to rid the world of plastic pollution. The coalition's website has numerous facts, news articles, and photos related to plastic pollution, as well as a link to its Last Plastic Straw project.

Sierra Club—www.sierraclub.org

Founded in 1892 by preservationist John Muir, the Sierra Club is a nonprofit organization that seeks to protect the environment and promote environmental policies. Its website has numerous articles on issues related to the environment, information on upcoming events, and tips on ways to help protect the planet.

Stop Food Waste—http://stopfoodwaste.org

This website features facts, videos, and blog posts related to food waste; tips on ways to help save food; a downloadable guide to help identify food waste in one's own home; and tips on keeping food fresh longer. The website also provides numerous links to external resources and organizations.

United Nations Environment Programme (UNEP)—
www.unenvironment.org

The UNEP coordinates the United Nations' environmental activities and helps developing countries implement environmentally sound policies and practices. The UNEP website contains reports, articles, and data related to protecting the planet, as well as information on upcoming events and ways to get involved.

US Environmental Protection Agency (EPA)—www.epa.gov

The EPA is a federal agency that is tasked with protecting human health and the nation's environment. Its website has a wealth of information on topics related to garbage and waste, including recycling, landfills, plastic pollution, food waste and recovery, electronic waste, and more.

For Further Research

Books

Michiel Roscam Abbing, *Plastic Soup: An Atlas of Ocean Pollution*. Washington, DC: Island, 2019.

David M. Barker, *E-Waste*. Edina, MN: ABDO, 2019.

Jen Chillingsworth, *Live Green: 52 Steps for a More Sustainable Life*. London: Quadrille, 2019.

Kathryn Kellogg, *101 Ways to Go Zero Waste*. Woodstock, VT: Countryman, 2019.

Laura Perdew, *The Great Pacific Garbage Patch*. Edina, MN: ABDO, 2018.

Beth Porter, *Reduce, Reuse, Reimagine: Sorting Out the Recycling System*. Lanham, MD: Rowman & Littlefield, 2018.

Adrienne Wheeler, *Waste Disposal*. New York: Rosen, 2018.

Philip Wolny, *The Fight for the Environment*. New York: Rosen, 2020.

Internet Sources

Elizabeth Balkan, "NRDC: A Holistic Approach to Reducing Food Waste," Natural Resources Defense Council, February 6, 2019. www.nrdc.org.

Rose Davidson, "Save the Earth!," National Geographic Kids. https://kids.nationalgeographic.com.

Dawn Gifford, "37 Ways to Reduce Trash in Your Home," *Small Footprint Family* (blog), 2017. www.smallfootprintfamily.com.

Idaho Public Television, "Garbage: Facts," 2019. http://idahoptv.org.

Arlene Karidis, "What Will the Future Landfill Look Like?," Waste360, July 5, 2018. www.waste360.com.

Tonya Mosley and Samantha Raphelson, "Exposing the Myth of Plastic Recycling: Why a Majority Is Burned or Thrown in a Landfill," WBUR, September 20, 2019. www.wbur.org.

Janelle Nanos, "I Broke Up with Plastic, and You Can, Too," *Boston Globe*, April 10, 2019. www.bostonglobe.com.

National Geographic, "7 Things You Didn't Know About Plastic (and Recycling)," April 4, 2018. https://blog.nationalgeographic.org.

Kirstie Pecci, "All Landfills Leak, and Our Health and Environment Pay the Toxic Price," Conservative Law Foundation, July 23, 2018. www.clf.org.

Plastic Pollution Coalition, "Take Action," 2019. www.plasticpollutioncoalition.org.

Brandon Pytel, "You're Doing It Wrong: 7 Tips to Recycle Better," Earth Day Network, October 1, 2019. www.earthday.org.

Bernhard Schroeder, "The Massive Industry of Food Waste Provides Several Opportunities for Creative Entrepreneurs," *Forbes*, October 2, 2019. www.forbes.com.

Alana Semuels, "The World Has an E-Waste Problem," *Time*, May 23, 2019. https://time.com.

Sierra Club, "What It's like Living in a Landfill," July 6, 2019. www.sierraclub.org.

Washington Post, "How Much Do You Know About Earth Day and the Environment?," September 25, 2019. www.washingtonpost.com.

Index

Picture Credits

Cover: Kanvag/Shutterstock

6: Maury Aaseng

11: Associated Press

14: Shutterstock.com

17: Shutterstock.com

22: rblfmr/Shutterstock

26: Mohamed Abdulrahheem/Shutterstock

33: zlikovek/Shutterstock

37: Aline Tong/Shutterstock

38: Parilov/Shutterstock

43: Oskari Heimonen/Shutterstock

46: David A. Liman/Shutterstock

49: EleniyaChe/Shutterstock

54: Art Babych/Shutterstock

57: Newscom Images

60: Pumidol/Shutterstock

About the Author

Peggy J. Parks has written more than 150 educational books on a wide variety of topics for students of all ages. She holds a bachelor's degree from Aquinas College in Grand Rapids, Michigan, where she graduated magna cum laude. Parks lives in Muskegon, Michigan, a town she says inspires her writing because of its location on the shores of beautiful Lake Michigan.